POSTMODERN ENCOUNTERS

Thomas Sebeok and the Signs of Life

Susan Petrilli and Augusto Ponzio

Series editor: Richard Appignanesi

ICON BOOKS UK

TOTEM BOOKS USA

Published in the UK in 2001
by Icon Books Ltd., Grange Road,
Duxford, Cambridge CB2 4QF
E-mail: info@iconbooks.co.uk
www.iconbooks.co.uk

Published in the USA in 2001
by Totem Books
Inquiries to: Icon Books Ltd.,
Grange Road, Duxford,
Cambridge CB2 4QF, UK

Sold in the UK, Europe, South Africa
and Asia by Faber and Faber Ltd.,
3 Queen Square, London WC1N 3AU
or their agents

Distributed to the trade in the USA by
National Book Network Inc.,
4720 Boston Way, Lanham,
Maryland 20706

Distributed in the UK, Europe,
South Africa and Asia by
Macmillan Distribution Ltd.,
Houndmills, Basingstoke RG21 6XS

Distributed in Canada by
Penguin Books Canada,
10 Alcorn Avenue, Suite 300,
Toronto, Ontario M4V 3B2

Published in Australia in 2001
by Allen & Unwin Pty. Ltd.,
83 Alexander Street,
Crows Nest, NSW 2065

Series editor: Richard Appignanesi

ISBN 1 84046 278 7

Typesetting by Wayzgoose

Printed and bound in the UK by
Cox & Wyman Ltd., Reading

The Life of Signs and the Signs of Life

Although it is not easy to imagine it now, the last century may become known in the future as the period in which signs were rediscovered. The deaths of Charles S. Peirce and Ferdinand de Saussure in 1914 and 1913 respectively, left the twentieth-century world a legacy of intellectual tools to take stock of a fundamental fact of existence: the sign. In the later part of the century, especially, it became clear that information had surpassed the value of all other traditional raw materials and commodities. Technologies such as the telephone, television, computers and the Internet electronically delivered the diversity of human sign production into the domestic environment. Truly, the final decades of the century could be called an 'epoch of signs'.

Concurrently, intellectual life responded in a concerted manner. From the 1960s onwards, 'semiotics', the study of the sign, began to transform the human sciences and, later, the natural sciences. In Europe, following the work of Saussure especially, sign study focused on the human use of signification, homing in on the logic of 'communication' and exposing the 'codes' underlying diverse cultural phenomena. The study of sign systems reinvigorated established disciplines and acted as midwife to new disciplines such as communications, media and cultural studies.

In the popular imagination, semiotics seemed to be the method by which one could study television, uncover the devices of photography, demystify literary texts, reveal the construction of films and systematically analyse numerous other features of popular culture. In truth, however, semiotics as an international discipline was not exclusively devoted to such parochial aims. Running through the entire semiotic enterprise was a much bolder conception of the universe of signs, a conception that went way beyond human sign use and which, ultimately, would be able to reveal to humans profound facts about their co-existence with other organisms and other worlds.

Much of this work had been proceeding for some decades, although it was to come to fruition especially in the 1990s and into the new century. Above all, the enterprise we have described is associated with the name of Thomas A. Sebeok.

Sebeok[1] is one of the scholars who has most contributed to establishing semiotics as a field and interdisciplinary perspective. His research is largely inspired by Charles S. Peirce (1839–1914), though his *maîtres à penser* also include such figures as Charles Morris (1901–79) and Roman Jakobson (1896–1982), whose work under certain aspects he continues.

Sebeok's interests cover a broad band of territories ranging from the natural sciences to the human sciences. Consequently, he deals with theoretical issues and their applications from as many angles as the disciplines he traverses: linguistics, cultural anthropology, psychology, artificial intelligence, zoology, ethology, biology, medicine, robotics, mathematics, philosophy, literature, narratology and so forth. Even though the initial impression might be that Sebeok's work proceeds rather erratically as he experiments with varying perspectives and embarks upon different research ventures, in reality his expansive and seemingly disparate interests find a focus in his 'doctrine of signs'. Indeed, the fundamental conviction underpinning his general method of enquiry is that the universe is perfused with signs and, as Peirce hazards, may be composed exclusively of signs.

Semiotics is thus the place where the 'life sciences' and the 'sign sciences' converge. This means that *signs* and *life* converge. Therefore, it follows that the human being is a sign in a universe of signs.

Sebeok extends the boundaries of traditional sign study, providing a 'semiotics' far more comprehensive than 'semiology'. The limit of 'semiology', Saussure's projected science of the sign, consists in the fact that it mistakes a part (that is, human signs and in particular verbal signs) for the whole (that is, all possible signs, human and non-human). On the basis of such a mystification, semiology incorrectly claims to be the general science of signs. When the general science of signs chooses the term 'semiotics' for itself, the aim is to underline the distance from semiology and its errors. Sebeok dubs the semiological tradition in the study of signs the 'minor tradition', and the tradition he promotes instead the 'major tradition', as represented by John Locke and Peirce, as well as studies on signs and symptoms by Hippocrates (*c.* 460–370 BC) and Galen (AD 129–*c.* 200). Semiotics, therefore, is at once recent – if considered from the viewpoint of the determination of its status and awareness of its wide-ranging possible applications; and it is also ancient – if its roots are traced back, following Sebeok,[2] to the theory and practice of ancient medicine.

Through his numerous publications Sebeok has propounded a wide-ranging vision of semiotics that coincides with the study of the evolution of life.

6

After Sebeok's work, both the conception of the semiotic field and history of semiotics are insuperably changed. Thanks to him, semiotics at the beginning of the new millennium has broad horizons; far broader than those envisaged by sign study in the first half of the 1960s.

In what may be defined, then, as a 'global' or 'holistic' approach to sign studies, Sebeok extends his gaze over the whole universe in so far as it teems with information, messages and signifying processes; a universe that is characterised, as anticipated, and as he never tires of repeating, as a fact of *signification* long before becoming a fact of *communication*.[3] Communication suggests understanding – even if limited – between a sender and a receiver; yet, signification can take place without such understanding and without the intention of conveying a message. As Sebeok playfully put it during a seminar held in 1987, 'Semiotic and Communication':

The world is composed entirely of signs, and therefore, I think of the whole world as my oyster; whereas for some people only the human world, and then only a small portion of that, is their oyster.[4]

In Sebeok's view, then, semiosis and life coincide.

This belief leads to an intriguing hypothesis: given that semiosis or sign behaviour involves the whole living universe, a full understanding of the dynamics of semiosis may in the last analysis lead to a definition of life itself. Semiosis originates with the first stirrings of life on the planet. This leads Sebeok to formulate an axiom that he believes is cardinal to semiotics: 'Semiosis is the criterial attribute of life'[5] – that is, 'the criterial mark of all life is semiosis' (where 'semiosis' means the activity of signs). His second axiom, 'semiosis presupposes life', complements the first.[6] It is hardly surprising, then, that all the life sciences find a place in Sebeok's intellectual horizon, estimated in their importance for a full understanding of signs and their workings in the terrestrial 'biosphere'.[7] Hence, 'global semiotics' provides a meeting point and an observation post for studies on the life of signs and the signs of life.[8]

In line with the 'major tradition' in semiotics, Sebeok's global approach to sign life presupposes his critique of anthropocentric and glottocentric semiotic theory and practice. In his explorations of the boundaries and margins of the science or 'doctrine' of signs,[9] Sebeok opens the field to include *zoosemiotics* (a term he introduced in 1963), or, even more broadly, *biosemiotics* on the one hand

and *endosemiotics* on the other. In Sebeok's conception, the sign science is not only the study of communication in culture or concerned with the social life of signs as proposed by Saussure but also the study of communicative behaviour in a biosemiotic perspective. Consequently, Sebeok's global semiotics is characterised by a maximum broadening of competencies.

Semiotics is Not Only Anthroposemiotics

Before contemplating the signs of unintentional communication (semiology of signification), semiotics was further limited by an exclusive concentration on the signs of intentional communication (semiology of communication). These were the main trends in semiology following Saussure. In Sebeok's conception, sign science not only studies communication in culture, but also communicative behaviour of a biosemiotic order – that is to say, biosemiotics is the wider context of all semiotics. He writes:

Biological foundations lie at the very epicenter of the study of both communication and signification in the human animal.[10]

Sign study will remain forever blinkered if it insists on linguistics or the signs used by humans as its fundamental model. Instead, it must look to the root biological processes that are at the foundation of all sign activity. Freeing oneself from the anthropocentric (or cultural) perspective as it has characterised semiotics from the 1960s to the 1980s implies taking into account other sign systems beyond those specific to mankind. That the sign systems of other living beings are not *alien* to the human world means there is a point of encounter between human communication and the communicative behaviour of non-human communities. Such sign systems also involve human communication with the environment, as well as the sphere of endosemiotics (the study of cybernetic systems inside the body). In turn, the study of such systems within the body can take place at the level of individual (ontogenetic) development and at the level of a species (phylogenetic development).

However, it must be noted that Sebeok's work succeeds in avoiding any form of biologism as occurs when human culture is reduced to communication systems that can be traced in other species. Conversely, he avoids the anthropomorphic reduction of non-human animal communication to characteristic traits and models specific to mankind. Consequently, his doctrine of signs insists particularly on the autonomy of non-verbal sign systems. In spite of the predominance of verbal language in the sphere of anthroposemiosis, his work is always acutely aware of human sign systems that depend on the verbal only in part.

In short, semiotics after Sebeok is not only *anthroposemiotics* but also *zoosemiotics*, *phytosemiotics*, *mycosemiotics*, *microsemiotics*, *endosemiotics*, *machine semiotics* and *environmental semiotics*, all under the umbrella of *biosemiotics* or, increasingly now and in the future, just plain *semiotics*.

A Transitional Book

In the opening lines to *The Sign & Its Masters*,[11] Sebeok describes this book of 1979 as 'transitional', being a remark that may be extended, in truth, to the whole of his research if considered in the light of recent developments in philosophico-linguistic and semiotic debate. However, our allusion is to the transition from 'code semiotics', which is centred on linguistics (and, therefore, verbal signs), to 'interpretation semiotics', which (unlike the former) also accounts for the autonomy and arbitrariness of non-verbal signs, whether 'cultural' (dance, semaphore) or 'natural' (such as signs among animals and plants).

In his survey of the problems relevant to semiotics and of the masters of signs, Sebeok discusses the various aspects characterising the 'cultural' and 'natural' approaches to semiotics, and which may be very simply summarised with two aforementioned names – Ferdinand de Saussure and Charles S. Peirce. The study of signs is 'in transit' from 'code semiotics' to 'interpretation semiotics' as represented by these two emblematic figures, and in fact has now decidedly shifted in the direction of the latter.

An earlier book of 1976, *Contributions to the Doctrine of Signs*, has a strong theoretical bias; and in it Sebeok had already expressed his preference for the semiotics of interpretation. *The Play of Musement*, a collection of papers published in 1981,[12] explores the efficacy of semiotics as a methodological tool and the potential range of its application and does so in more discursive terms, although in both these books Sebeok's perspective has solid theoretical foundations.

By contrast, *The Sign & Its Masters*, the in-between book, considers the different possibilities that branch out from our two semiotic alternatives, code semiotics and interpretation semiotics. In fact, in addition to being a compact theoretical book, *The Sign & Its Masters* also offers a survey of the various alternatives, positions and phases in sign studies as they have been incarnated through history by important scholars of signs who have dealt with signs either directly or indirectly.

Sebeok's writings transform us into the direct witnesses of turning points in his research as he experiments, discusses and evaluates different methods of semiotic inquiry, identifies possible objects of analysis and outlines the boundaries, or, better, suggests the boundlessness of semiotics as a disciplinary field. From this point of view, *The Sign & Its Masters* – as, in reality, all of his research – is *transitional* in so far as it contributes significantly to the shift towards interpretation semiotics. This shift frees sign study once and for all from subordination to (Saussurean) linguistics and from false dichotomies: communication semiotics versus signification semiotics, referential semantics versus nonreferential semantics.[13]

I Think I Am a Verb of 1986 is the fourth book in Sebeok's tetralogy of the 1970s and 1980s. Since then other important volumes have followed in rapid succession. These include: *Essays in Zoosemiotics* (1990), *A Sign is Just a Sign* (1991),

American Signatures (1991), *Semiotics in the United States* (1991), *Signs: An Introduction to Semiotics* (1994), *Come comunicano gli animali che non parlano* (1998) and *Global Semiotics* (2001),[14] without forgetting important earlier volumes such as *Perspectives in Zoosemiotics* (1972), plus numerous others under his editorship including *Animal Communication* (1968), *Sight, Sound, and Sense* (1978), and *How Animals Communicate* (1979).[15]

Rather than continue this long list of publications, it will suffice to remember that Sebeok has been publishing since 1942. His writings are the expression of ongoing research and probing reflection over more than half a century as he interprets the semiosic universe, whose infinite multiplicity, variety and articulation he has substantially contributed to making manifest.

I Think I Am a Verb is a book that at once assembles a broad range of interests and that also acts as a launching pad for new research itineraries in the vast region of semiotics. The title evokes the dying words of the eighteenth President of the United States, Ulysses Grant, which ring with Peircean overtones. In fact, in Peirce's view, man is a sign just as all living beings are. However, Sebeok's choice of a verb instead of a noun to characterise this sign serves to emphasise the condition of continuous becoming, transformation and renewal of signs in the human world.

A fundamental point in Sebeok's doctrine of signs is that living is sign activity. To maintain and to reproduce life, and not only to interpret it at a scientific level, are all activities that necessarily involve the use of signs. Sebeok theorises a direct connection between the biological and the semiosic universes, and, therefore, between biology and semiotics. His research would seem to develop Peirce's conviction that man is a sign with the addition that this sign is a verb: to interpret. And in Sebeok's particular conception of reality, the interpreting activity coincides with the life activity, and in his own personal case, with the whole of his life. If I am a sign, as he would seem to be saying through his life as a researcher, then nothing that is a sign is alien to me – *nihil signi mihi alienum puto*. And if the sign situated in the interminable chain of signs is necessarily an 'interpretant' – the term Peirce gave to the effect of a sign, an effect that is itself a sign – then 'to *interpret*' is the verb that may best help me understand who I am.

Sebeok's position is distant from Saussure's, who limited the sign science to the narrow spaces of the signs of human culture and, still more reductively, to signs produced intentionally for communication. Instead, for Sebeok, no aspect of sign life must be excluded, just as no limits are acceptable on semiotics, whether contingent (for example, political factors) or deriving from epistemological conviction (a bias towards one particular theory of knowledge). Contrary to first impressions, though, Sebeok's work does not claim the status of scientific or philosophical omniscience, or the ability to solve all problems indiscriminately.

We believe that Sebeok's awareness of the vastness, variety and complexity of the territories he is committed to exploring and of the problems he analyses, demonstrates a sense of utmost prudence, sensitivity to problems and humility in the interpretations he offers. This is the case not just when he is venturing over the treacherous territory of signs, but still more in relation to the deceptive sphere of the signs of signs – the place of his semiotic probings.

Sebeok's Semiotic Research

Sebeok began his studies in higher education during the second half of the 1930s at Cambridge. He was particularly influenced by *The Meaning of Meaning* (1923) by Charles K. Ogden and Ivor A. Richards long before it became a classic in semiotics.[16] Also, he can boast of having benefited from direct contacts with two great masters of the sign mentioned earlier who, in different ways, and under different aspects were also his teachers: Charles Morris and Roman Jakobson.[17]

Let us now list and at once distinguish between the various aspects and parts of the multifarious 'semiosic universe' as it emerges from Sebeok's semiotic research.

In Sebeok's view, the universe is perfused with signs. These signs are interconnected and interdependent, and form a huge semiosic 'web' – to use an image lauched by Sebeok in 1975. Sign science or semiotics is the place where studies on the life of signs and on the signs of life converge. Through his analyses of the signifying material making up the biosphere (the sphere of life), Sebeok contemplates the whole universe, *à la* Peirce, as *itself* a sign:

a vast representamen, a great symbol ... an argument ... necessarily a great work of art, a great poem ... a symphony ... a painting.[18]

Sebeok turns his attention to signs that are commonly the object of study by specialists from different fields, viewing them at once in their specificity and interrelatedness. These signs range from the signs of 'nature' to the signs of 'culture', from human signs to animal signs, from verbal signs to non-verbal signs, from natural languages to artificial languages, from highly 'plurivocal' and 'dialogic' signs to 'univocal' and 'monological' signs or, better, signals, from the signs of conscious life to the signs of unconscious life. These signs are endowed with varying degrees of 'indexicality' (a relation of causality and/or contiguity between a sign and that which it signifies – for example, a symptom on the skin caused by a virus in the body); 'symbolicity' (arbitrariness between sign and object, such as using the word 'beer' in English to refer to the class of drink called 'cerveza' in Spanish); and 'iconicity' (a relation of resemblance as in the case of a portrait, a diagram, a metaphor, a translation).

Looking to the whole universe, Sebeok's expansive gaze is the sign of his profound awareness that signs are interdependent and relational as he demonstrates how an understanding of any one particular type of sign – such as the verbal – is only possible in the light of its relation with other signs in the great sign network. In Sebeok's ecumenical perspective, therefore, the signs of nature and of culture forming this network are not considered as divided and separate but as interpretants, 'significate effects', of each other.

With reference to this last point and in debate with major exponents representing different trends in semiotics today, Sebeok states that:

[T]o me ... the imperium of Nature, or Weltbuch, over Culture, or Bucherwelt, has always been unmistakable. Only a patent theoretical basis was veiled to resolve what Blumenberg[19] has called an 'alte Feindschaft' between these two semiotic systems, the latter obviously immersed in the former. This is why my 'rediscovery' of the Umweltlehre came as such a personal revelation.[20]

Sebeok is here referencing the work of one of his masters, Jakob von Uexküll (1864–1944). For Uexküll, every organism exists in the world specific to its own species, which he calls an 'Umwelt'. Umwelten may be either simple as in the case of primitive organisms such as protozoa, or complex as in the more developed organisms. Particularly complex is the human Umwelt (in reality Umwelten because as we shall see later, human beings can have more than one world), which even foresees space and time co-ordinates.

The numerous human worlds, whether real or possible, are part of what the Russian semiotician Jurij M. Lotman (1922–93) calls the 'semiosphere' (that is, the totality of all human signs). A strong temptation for semiotics is to make all signs coincide with the signs of the semiosphere instead of recognising that the semiosphere limited to human culture is only a part of the biosphere. In truth, the real semiosphere – understood as the sphere of all signs – is the biosemiosphere. Therefore, Sebeok's conviction is that the anthropocentric approach to signs is the result of an understandably short-sighted view that limits signs to the species-specific human 'cultural' world. But 'culture' cannot be separated from 'nature' because it is a part of it. All the same, there is no doubt that the inner human world, with great effort and serious study, may reach an understanding of non-human worlds and of its connection with them.

Sebeok's Semiosic Universe

Sebeok's semiosic universe thus comprises the following.

• The life of signs and the signs of life as they appear today in the biological sciences: the signs of animal life and of specifically human life, the signs of adult life, and of the organism's relations with the environment, the signs of normal or pathological forms of dissolution and deterioration of communicative capabilities.

• Human verbal and non-verbal signs. Human non-verbal signs include signs dependent on natural language (as opposed to 'artificial', 'technical', 'specialist' language; natural languages such as English, German, Spanish, Chinese and so on are obviously also historico-social languages), and signs which, on the contrary, are not dependent on natural language and which, therefore, exist beyond the categories of linguistics. These include the signs of 'parasitic' languages such as artificial languages (Esperanto and so on), the signs of 'gestural languages' such as the sign language of Amerindian and Australian aborigines,[21] and the language of deaf-mutes, the signs of infants and the signs of the human body both in its more culturally dependent and in its natural-biological manifestations.

• Human intentional signs controlled by the will, and unintentional, unconscious signs such as those that pass in communication between human beings and animals in 'Clever Hans' cases.[22] Here, animals seem capable of certain performances (for example, counting) simply because they respond to unintentional and involuntary suggestions from their trainers. This group includes signs at all levels of conscious and unconscious life, and signs in all forms of lying (which Sebeok identifies and studies in animals as well), deceit, self-deceit and good faith.

• Signs at a maximum degree of plurivocality (that is, signs with a number of different possible meanings and therefore interpretations), and, on the contrary, signs that are characterised by univocality and which, therefore, are signals (that is, a sign that triggers just one reaction, such as the starting pistol in a race).

• Signs viewed in all their shadings of indexicality (as stated above: relationship of causality and/or contiguity between a sign and that which it signifies), symbolicity (relationship of arbitrariness between sign and object), and iconicity (relationship of resemblance).[23]

• Finally, 'signs of the masters of signs'. Those

through which it is possible to trace the origins of semiotics (for example, in its ancient relation to divination and to medicine), or through which we may identify the scholars who have contributed directly or indirectly (as 'cryptosemioticians') to the characterisation and development of this science, or 'signs of the masters of signs' through which we may establish the origins and development of semiotics relative to a given nation or culture, as in Sebeok's study on semiotics in the United States. 'Signs of the masters of signs' also include the narrative signs of anecdotes, testimonies and personal memoirs that reveal these masters not only as scholars but also as persons – their character, behaviour, everyday habits. Not even these signs, 'human, too human', escape Sebeok's semiotic interests.

All this is a far cry from the limited science of signs as conceived in the Saussurean tradition!

Metascience and 'Doctrine of Signs'

Sebeok's semiotics unites what other fields of knowledge and human praxis generally keep separate either for justified exigencies of a specialised order, or because of a useless, even harmful tendency towards short-sighted sectorialisation. Such an attitude is not free of ideological implications, which are often poorly masked by motivations of a scientific order.

Biology and the social sciences, ethology and linguistics, psychology and the health sciences, their internal specialisations – from genetics to medical semiotics, psychoanalysis, gerontology and immunology – all find in semiotics, as conceived by Sebeok, the place of encounter and reciprocal exchange, as well as of systematisation and unification.

At the same time, it must be stressed that systematisation and unification are not understood here in the static terms of an 'encyclopaedia', whether this takes the form of the juxtaposition of knowledge and linguistic practices or of the reduction of knowledge to a single scientific field and its relative language.

Global semiotics may be presented as a *metascience* that takes all sign-related academic disciplines as its field. It cannot be reduced to the status of philosophy of science, although as a science it is engaged in dialogic exchange with philosophy.

Sebeok achieves a global view through a continuous and creative shift in perspective that favours the development of new interdisciplinary relationships and new interpretive practices. Sign relations are identified where, for some, there seemed to be no more than mere 'facts' and relations among things, independent from communicative and interpretive processes. Moreover, this continual shift in perspective also favours the discovery of new cognitive fields and languages, which act dialogically. They are interpreted as signs and, as such, the act of interpretation is an 'interpretant' which, in the Peircean schema, although an 'effect' of the sign, can itself become a new sign. As he explores the boundaries and margins of the sciences, Sebeok dubs this open nature of semiotics the 'doctrine of signs'.

Semiotics as the 'Doctrine of Signs'

Despite such a totalising orientation it is notable that Sebeok uses neither the ennobling term 'science' nor the term 'theory' to name it. Instead, as we have seen, he repeatedly favours the expression 'doctrine of signs', adapted from John Locke according to whom a doctrine is a body of principles and opinions that vaguely form a field of knowledge. Sebeok also uses this expression as understood by Charles S. Peirce (that is, with reference to the instances of Kantian critique). This is to say that Sebeok invests semiotics not only with the task of observing and describing phenomena, in this case signs, but also of interrogating the conditions of possibility that characterise and specify signs for what they are, as they emerge from observation (necessarily limited and partial), and for what they must be.[24]

This humble and at the same time ambitious character of the 'doctrine of signs' leads Sebeok to a Kantian critical interrogation of its very conditions of possibility: the doctrine of signs is the sign science that questions itself, attempts to answer for itself, and inquires into its very own foundations. As a doctrine of signs, semiotics is also philosophy not because it deludes itself into believing it can substitute philosophy, but because it *does not* delude itself into believing that the study of signs is possible without philosophical questions regarding its conditions of possibility.

How is Semiotics as a Science and Metascience Possible?

Sebeok most significantly adds another meaning to 'semiotics' beyond the general science of signs: as indicating, that is, *the specificity of human semiosis*. This concept is clearly proposed in a paper of 1989, 'Semiosis and Semiotics: What Lies In Their Future?',[25] and is of vital importance for a *transcendental founding of semiotics* given that it explains how semiotics as a science and metascience is possible. He writes:

Semiotics is an exclusively human style of inquiry, consisting of the contemplation – whether informally or in formalised fashion – of semiosis. This search will, it is safe to predict, continue at least as long as our genus survives, much as it has existed, for about three million years, in the successive expressions of Homo, *variously labelled – reflecting, among other attributes, a growth in brain capacity with con- comitant cognitive abilities –* habilis, erectus, sapiens, neanderthalensis, *and now* s. sapiens. *Semiotics, in other words, simply points to the universal propen- sity of the human mind for reverie focused specularly inward upon its own long-term cognitive strategy and daily manoeuvrings. Locke designated this quest as a search for 'humane understanding'; Peirce, as 'the play of musement'.*[26]

This use of the term semiotics encapsulates the idea of the general study of signs and of the typology of semiosis (sign activity). In his article 'The Evolution of Semiosis',[27] Sebeok explains the correspondences that exist between the branches of semiotics and the different types of semiosis, from the world of micro- organisms to big kingdoms and the human world. Specific human semiosis, anthroposemiosis, is rep- resented as semiotics thanks to a 'modelling' device specific to humans, called by Sebeok 'language'. This observation is based on the fact that it is virtually certain that *Homo habilis* (about two million years ago) was endowed with language, but not speech. That is to say, humans possessed the capacity for *language* as a cognitive device for differentiating long before they started to implement it through *speech* for the purposes of verbal *communication*. Prior to the verbal form, *communication* would have taken place by non-verbal means.[28]

In the world of life, which coincides with semiosis,[29] human semiosis is characterised as metasemiosis. In other words, human semiosis offers the possibility of reflecting on signs, of making signs the object of interpretation not only in terms of the response to signs, but also of reflection on signs, where response is suspended and deliberation is possible. This exquisitely specific human capacity for metasemiosis may also be called 'semiotics'.

Developing Aristotle's (384–322 BC) astute observation made at the beginning of his influential philosophical work *Metaphysics*, that man tends by nature to knowledge, we might add that man tends by nature to semiotics. Human semiosis characteristically presents itself as *semiotics*.

Semiotics as human semiosis or anthroposemiosis can:

• either venture as far as the entire universe in search of meanings and senses, considering it therefore from the viewpoint of signs

• or absolutise anthroposemiosis by identifying it with semiosis itself.

In the case of the former, semiotics as a discipline or science (Saussure) or theory (Morris) or doctrine (Sebeok), presents itself as 'global semiotics' (Sebeok) and can be extended to the whole universe in so far as the universe is perfused with signs (Peirce). In the case of the latter, on the other hand, semiotics is limited and anthropocentric.

Three Aspects of the Unifying Function of Semiotics

As emerges in Sebeok's research, the unifying function of semiotics may be considered from the viewpoint of three strictly interrelated aspects all belonging to the same interpretive practice.

First, we always interpret: not only when we reason but also when all we wish to know is *what* or *how* we feel (hot, cold, fear, appetite, love, hatred, boredom and so on). Each time we become aware of a sensation, even the most immediate, we are carrying out an interpretation. Furthermore, to interpret means to form hypotheses and, therefore, to risk error. Unfortunately, our interpretations cannot refer to fixed and unquestionable rules as occurs, for example, by way of exception in mathematical demonstration (deduction). Nor is it possible to ascertain that what we hypothesise is always the case such as to formulate general rules on the basis of a complete test of *all* specific cases (induction) – there is simply no time for this in people's lives.

How, then, do we interpret? By guessing. This is a simple reply, but we had to wait for Peirce before it was formulated. Peirce calls this kind of interpretive process 'abduction' (as distinct from deduction and induction). With the word abduction or 'retroduction', Peirce indicates an interpretive process without pre-established rules that guarantee it. Indeed, abductive reasoning begins from specific cases and hazards the hypothesis of a general rule that is able to explain them, and that must be discovered or invented by guessing. The rule is true only on this condition that it be confirmed by the case in question. Human interpretive processes are mostly of the abductive type, and the more risky the guessing game the more these processes are innovative and creative. Interpretations by semiotics are the expression of high level abductive reasoning.

We shall now consider how Sebeok uses interpretive practices of the abductive type to identify signs and relations among signs. From this point of view we must underline three aspects of semiotic analysis, which are covered in further detail below.

1. The descriptive-explanatory aspect.
2. The methodological aspect.
3. The ethical aspect.

1. The descriptive-explanatory aspect

Semiotics singles out, describes and explains signs forming events that may be considered as signs of one another. Something is a sign of something else for three fundamental reasons.

• The sign and that of which it is a sign may be connected by a relation of contiguity and causality (smoke and fire, cloud and rain): in this case the relation is 'indexical' and the sign is an 'index'.

• Or the sign and that of which it is a sign may be associated on the basis of a relation of resemblance (which may be vague or questionable) rather than by factors of contiguity and causality ('life' and 'waltz' in the metaphor 'life is a waltz'): in this case the relation is 'iconic' and the sign is an 'icon'.

• Or the sign and that of which it is a sign may be connected by *social convention* rather than by factors of contiguity and causality or of similarity ('water' in English, 'acqua' in Italian for H_2O in the language of chemistry): in this case the relation is 'symbolic' and the sign is a 'symbol'.

Semiotics from the viewpoint of its descriptive-explanatory aspect evidences sign relations even where they have not been observed. It does so not only in relation to objects, events and phenomena, but also in relation to the sciences that study these things. For example, through interpretations that are highly abductive (that is, hypothetical and risky), semiotics establishes a connection between its own studies and biology. Thanks to the capacity for innovation as determined in abductive interpretive processes, semiotics, which consequently knows how to look for resemblances among fields that would seem to be completely different, and that is interested in signs wherever they occur, succeeds in linking scientific fields even when they would seem to be distant from each other.

The relation of resemblance or similarity identified by semiotics is very different from the immediate and superficial type of resemblance called 'analogy' in the language of biology as distinct from resemblances of the genetical-structural order called instead 'homology'. In comparative biology, resemblance between the wing of a bird and the wing of an insect is not significant, even though they are both called wings. Instead, of great interest in scientific terms is resemblance between things that would seem very different, such as the fin of a fish, a bird's wing and the human limb. As anticipated, this kind of resemblance is of the genetic and structural order (that is, pertaining to the sphere of homology). When semiotics states that 'life is semiosis' (that is, sign activity) it is establishing a relation of similarity between life and signs of the homological type. These relations are in fact completely different from superficial likenesses, and therefore from simple analogies, such as to be able to state that 'life is a waltz'.

2. The methodological aspect

Semiotics is also the search for methods of inquiry and acquisition of knowledge, both ordinary and scientific knowledge. From this point of view, and differently from the first aspect, semiotics does not limit itself simply to describing and explaining, but it also makes proposals concerning cognitive behaviour. Under this aspect as well, then, semiotics overcomes the tendency to parochial specialism among the sciences when this causes separation from each other.

3. The ethical aspect

For the ethical aspect, we propose the term 'ethosemiotics' or 'telosemiotics', from 'telos' (end). Under this aspect, the unifying function of semiotics concerns proposals and practical orientations for human life in its wholeness (human life considered in all its biological and socio-cultural aspects). The focus is on what may be called the 'problem of happiness'. This problem is evidently considered to be very important by Herodotus (*c.* 484–*c.* 424 BC), the father of history, who early in the first book of the *Histories* narrates the downfall of the last of the Lydian kings, Croesus, who imagined himself to be the happiest of men.

In turn, the story of Croesus as described by Herodotus is interpreted by Sebeok: happiness is impossible for Croesus to maintain because of his inability to hold in due account the worlds (and signs) of each of his two sons: one endowed with the word, the other deaf and dumb, and unnamed. By a reversal of fortune, the silent, nameless son succeeds to speak and he saves his father's life; on the other hand, the able son is reduced to the silence of death by his eloquence. According to Sebeok, the moral of this story is that happiness requires that the flow of communication from the non-verbal to the verbal be restored, and that the consequences of blocking this flow are tragic.

Sebeok's study, 'The Two Sons of Croesus: A Myth about Communication in Herodotus',[30] reflects on this third aspect of semiotics which refers to the problem of wisdom as deposited in myths, popular tradition and literature in particular genres (those described by Mikhail Bakhtin as belonging to 'carnivalised literature' that derive from comic popular culture).[31] By analogy with the deaf and dumb son of Croesus, we may remember King Lear's reticent Cordelia, or in *The Merchant of Venice*, the muteness and simplicity of the leaden casket – contrary to common expectation a sign that it holds Portia's image.

Concerning this third aspect of the unifying function of semiotics, particular attention is paid to recovering the connection with what is considered and experienced as being separate. In today's world, the logic of production and the rules that govern the market allowing all to be exchanged and commodified, threaten to render humanity ever more insensible. Humans increasingly pay little or no attention to the signs of all that which cannot be measured or purchased but that are received as a gift (friendship, love, mercy, forgiveness, and the gift of life itself), and that actually play a major role in our lives. These signs may range from the vital signs forming the body to the seemingly futile signs of phatic communication with others (for example, the use of courtesy formulae or comments on the weather). Reconsideration of these signs and their relative interrelations would seem absolutely necessary in the present age to improve the quality of life. The economics of capitalist globalisation imposes ecological conditions in which communication between ourselves and our bodies, as well as the environment, is ever more difficult and distorted. Our bodies become self-centred separate entities trained to be exploited rather than to appreciate and increase the value of human existence and of life in general.[32]

Moreover, this third aspect of semiotics also operates in a way that unites rational world views to myth, legend, fable and all other forms of popular tradition on the relationship of humans to the world about them. Such a function is rich with implications for human behaviour: those signs of life that today we cannot or do not wish to read, or those signs that we do not know how to read, may well recover one day their importance and relevance for humanity.

Origin of Signs and Origin of Life

As the study of all kinds of messages, semiotics discovers 'semiosic events' in living organisms. Consequently, as mentioned, on the basis of interpretations, both inductive (that is, confirmed by many and different cases) and abductive (that is, based on hypotheses and considered valid until proven to the contrary), Sebeok believes that semiosic processes and life coincide. Such identification implies a correspondence between semiotics and *biosemiotics*, and invests *zoosemiotics* (with respect to which anthroposemiotics is a part) with a completely different role from that conceived by Umberto Eco[33] (1975) when he describes it as 'the inferior threshold of semiotics'. In the same way, interpretation of biosemiotics as a mere 'sector' of semiotics is also reductive.

In Sebeok's research, semiotics is interpreted and practised as a life science, as biosemiotics. It follows that his approach to semiotics may be situated in the tradition of thought established by the founders and masters of semiotics by such figures as those we have already named: Hippocrates, Galen, Peirce, von Uexküll and, in recent times, René Thom (an important topologist and Peirce scholar well-versed in the principles of biology).

In this context, Sebeok's semiotics examines the problem of the origin of signs, which is nothing less than the problem of the genesis of the universe. He deals with all types of semiosis, from the free flow of energy-information to signals and signs.

The development of semiosis and its complex articulation coincide with the evolution of terrestrial life from a single cell to its present-day multiform diversity, subdivided into three (or four) big cellular kingdoms: plants, animals and fungi. These kingdoms co-exist and interact with the microcosm and together they form the 'biosphere' or sphere of life; along with Lotman's 'semiosphere', as we have seen, they form the 'biosemiosphere'.

A characteristic trait of *human* semiosis is the

presence of verbal signs. However, to avoid interpretations of an anthropocentric or phonocentric
order (that is, biased towards humans and biased
towards the voice and *verbal signs*), human semiosis or anthroposemiosis must be considered in the
broader context of general semiosis (that is,
biosemiosis). As Sebeok states,[34] all terrestial life
functions through non-verbal signs, whereas only
human life functions through two types of signs –
verbal and non-verbal.

To Live and To Lie

In Italy, long before Eco defined semiotics as the
discipline that studies lying,[35] Giovanni Vailati
(1863–1909) had stated that signs may be used for
deviating and deceiving. He entitled his review of
Giuseppe Prezzolini's *L'arte di persuadere*, 'Un
manuale per bugiardi' ('A handbook for liars').[36]
This particular aspect of Vailati's studies is analysed
by Augusto Ponzio in his 1988 monograph on the
Italian philosopher and semiotician Ferruccio
Rossi-Landi (1921–85). 'Plurivocità, omologia,
menzogna' ('Plurivocality, homology, lying') is the
title of a section included in a chapter dedicated to
the relation between Rossi-Landi and Vailati,[37] his
predecessor. Sebeok himself also evokes Vailati in

relation to Peirce in his paper 'Peirce in Italia' of 1981.[38] He too analyses the use of signs for lying (a theme that forms yet another leitmotif in his research) – that is to say, the use of signs for fraud, illusion and deception, the capacity of signs for masking and pretence.

Deception, lying and illusion are forms of behaviour that a semiotician like Sebeok, entranced by signs wherever they occur, cannot resist. For example, he is attracted by the signs of the magician and constantly returns to forms of behaviour and situations reminiscent of Clever Hans, the horse that presumably knew how to read and write, but that in reality was an able interpreter of the signals communicated to it by its trainer, either inadvertently or voluntarily through an intentional attempt at fraud.[39]

Sebeok explores the capacity for lying in the non-human animal world, an interest we believe that has a double motivation. The first concerns his commitment to contradicting the belief that animals can 'talk' in a literal sense, that they are invested with a characteristic – language – that is species-specific exclusively to humankind. In some cases, this involves unmasking the fraudulent acts of impostors; in others it involves undermining illusions. Through theoretical discussion, documentation and even parody,[40] Sebeok has made an important contribution to evidencing the absurd, often ridiculous and no doubt scientifically unsound consequences of ignoring and abstracting from species-specific differences between human verbal communication and animal communication.

The second motivation is related to Sebeok's wish to explore the fascinating question of whether non-human animals lie, as humans do. As evidenced by studies in zoosemiotics, signs do not belong exclusively to the human world and it may well be that the use of signs also implies the ability to lie.[41]

Semiosic Excess Beyond Sign Function

The world of signs is obviously not only the world of deception but also of other practices (no doubt connected with it), such as playing, using symbols and making gifts. The fact that animals use signs implies that these practices as well, which are mostly considered as the prerogative of 'culture', may be traced in the non-human animal world. By contrast with those researchers who often insist emphatically or exclusively on the *function* of signs for the purposes of understanding the *nature* of signs, Sebeok highlights the importance of sign activity as an end in itself. In other words, sign activity takes place irrespective of specific functions and purposes, and therefore it is sometimes necessary to consider sign activity as a sort of idle, non-functional and unproductive semiotic mechanism.

Nor is this particular aspect of semiosis merely restricted to the signs of ritual behaviour in both human and non-human animals, and which as such do not have a discernible function. Verbal language – which, more often than not, is interpreted in relation to communicative function – is also better understood in terms of play and of the human propensity for fantasising and daydreaming or 'musement'.[42] The human propensity for musement implies the ability to carry out such operations as predicting the future or 'travelling' through the past – the ability, that is, to construct, deconstruct and reconstruct reality, thereby inventing new worlds and interpretive models. Let us remember that in interpreting Peirce, Sebeok borrows the happy expression *The Play of Musement* as the title of his book of 1981.

Indeed, as Peirce had already demonstrated, the capacity for inferential mechanisms that allow for the qualitative development of knowledge is fundamental to play and fantasy, as well as to the practices of inquiry and simulation. We are alluding to what Peirce calls 'abduction', or 'hypothesis', or 'guessing'. Surprising abductions are achieved in science with scientific discoveries, but also in police investigations where the astonishing solution of a case might depend on formulations at first sight risky. What the famous investigators Auguste Dupin and Sherlock Holmes (created by Edgar Allan Poe and Arthur Conan Doyle respectively and therefore by their 'play of musement') call 'analysis' or 'deduction' is in reality abduction.[43]

Semiotics itself is thus engaged in the play of musement. In the words of Sebeok:

The central preoccupation of semiotics is an illimitable array of concordant illusions; its main mission to mediate between reality and illusion – to reveal the substratal illusion underlying reality and to search for the reality that may, after all, lurk behind that illusion. This abductive assignment becomes, henceforth, the privilege of future generations to pursue, in so far as young people can be induced to heed the advice of their elected medicine men.[44]

And to show how the unconscious aspect of sign behaviour transcends the intentional symbolic order that is precisely orientated to functions and ends, in his text on 'The Two Sons of Croesus'[45] Sebeok refers to the problem of dreaming as well, this 'imaging process of communion', which he describes as 'silent, but eloquent nonetheless', the same process designated by the father of psychoanalysis Sigmund Freud as 'oneiric work'.[46]

The lack of functionality or forms of 'unproductive consumption' or of dissipation are identified by Sebeok as 'entropic' phases necessary to the development of life on earth. It is as though life is in continual need of – indeed is founded on – death, in order to reproduce and maintain itself. The implications of such a statement made within different trends in the history of philosophy are numerous, for what concerns sign theory is the implication that the semiotic chain is subject to loss, gaps, the erasing of sense. All this implies that in relation to sign material we must also necessarily postulate a sort of anti-material.

Sebeok points out the limitations of research on the nature of signs when it restricts its attention merely to sign *function*. Instead, he emphasises the importance of sign activity that is not directed towards precise goals and ends. The propensity for non-functional and unproductive sign activity is visible in ritual behaviour among human beings and animals, but also in verbal communication. In fact, beyond its communicative function, verbal expressions may be considered in terms of play without which imagination, fantasy or abductive reasoning at the highest degrees of innovation and invention would never have been possible.[47]

Modelling Systems Theory

A fundamental notion in Sebeok's semiotics is that of *model*. Sebeok in fact develops the concept of modelling as proposed by the so-called Moscow-Tartu school (A. A. Zaliznjak, V. V. Ivanov, V. N. Toporov and J. M. Lotman). For this school of semioticians, modelling is used to denote natural language ('primary modelling system') and the other human cultural systems ('secondary modelling systems' – such as the novel, painting, television and so on). Sebeok seeks to extend this concept beyond the domain of anthroposemiotics. By connecting 'modelling' with the biologist Jakob von Uexküll's concept of *Umwelt*, Sebeok's interpretation may be translated as an 'outside world model'. On the basis of research in biosemiotics, Sebeok holds that the modelling capacity is observable in all life forms.[48]

These terms need some elucidation. The study of modelling behaviour in and across all life forms requires a methodological framework developed from the field of biosemiotics. This methodological framework is *the modelling systems theory* proposed by Sebeok in his research on the interface between semiotics and biology. Modelling systems theory studies semiotic phenomena as modelling processes.[49]

In the light of semiotics viewed as a modelling systems theory, semiosis – a capacity of all life forms – may be defined as 'the capacity of a species to produce and comprehend the specific types of models it requires for processing and codifying perceptual input in its own way'.[50]

The applied study of modelling systems theory is called *systems analysis*, which distinguishes between primary, secondary and tertiary modelling systems.

The primary modelling system is the innate capacity for *simulative* modelling – in other words, it is a system that allows organisms to simulate something in species-specific ways (that is, in ways that are not available to another species).[51] Sebeok calls 'language' the species-specific primary modelling system of the species called *Homo*.

The secondary modelling system is one that

includes both 'indicational' and 'extensional' modelling processes. The non-verbal form of indicational modelling has been documented in various species: humans certainly have it, but so, too, do other animals. Extensional modelling, on the other hand, is a uniquely human capacity, because it presupposes *language* (primary modelling system) which, as we have seen, Sebeok distinguishes from *speech* (human secondary modelling system).[52]

The tertiary modelling system is one that undergirds highly abstract, symbol-based modelling processes. Tertiary modelling systems are the human cultural systems (painting, the novel, television and so on again) which the Moscow-Tartu school had mistakenly dubbed 'secondary' as a result of conflating 'speech' and 'language'.[53]

The Question of the Origin of Human Verbal Language

The question of the origin of human verbal language has often been set aside by the scientific community as unworthy of discussion, having most often given rise to statements that are naïve and unfounded.[54] Despite this general attitude, however, Sebeok neither forgets the problem of origins nor underestimates its importance.

He claims that human verbal language is species-specific. It is on this basis that he debates, often with great irony, the enthusiastic supporters of projects developed for teaching words to captive primates. Such bizarre behaviour is based on the false assumption that animals might be able to talk or, even more preposterous, that they possess the capacity for language. Sebeok's distinction between 'language' and 'speech'[55] not only protects against wrong-headed conclusions regarding animal communication, it also constitutes a general critique of the impulse to privilege the voice (phonocentrism) and to base scientific investigations on thoroughly human principles (anthropocentrism).[56]

According to Sebeok, language appeared and evolved as an *adaptation* much earlier than speech in the evolution of the human species to *Homo sapiens*. Language is not a communicative device (a point on which Sebeok is in accord with Noam Chomsky, even though the latter does not make the same distinction between *language* and *speech*); in other words, the specific function of language is not to transmit messages or to give information.

Sebeok instead describes language as a *modelling device*.[57] Every species is endowed with a model that 'produces' its own world, and language is the model belonging to human beings. However, as a modelling device, human language is completely different from the modelling devices of other life forms. Its characteristic trait is what the linguists call *syntax*, the ordering and operational rules of individual elements. But, while for linguists such elements ordered by syntax are words and phrases, Sebeok instead refers to a mute syntax when he speaks of syntax in language. This syntax orders the events and objects of human experience, transforming them into elements of its *Umwelt*. Through syntax, humans established operational rules to organise their relations with each other and the surrounding environment, even *before* speaking. As early as the hominids, syntax made it possible not only to have a 'reality' (that is, a world), but also to frame an indefinite number of possible worlds, this being a capacity that is unique to the human species.

Thanks to syntax, human language is like Lego building blocks. It can reassemble a limited number of construction pieces in an infinite number of different ways. As a modelling device, language can produce an indefinite number of models; in other words, the same pieces can be taken apart and put together to construct an infinite number of different models.

And thanks to language, not only do human animals produce worlds similarly to other species, but also, as the German philosopher Leibniz said, human beings can produce an infinite number of possible worlds. This brings us back to the 'play of musement', a human capacity that Sebeok following Peirce considers particularly important for scientific research and all forms of investigation, and not only for fiction and all forms of artistic creation.

Speech, like language, made its appearance as an adaptation, but *for the sake of communication* and much later than language, precisely with *Homo sapiens*. Consequently, language too ended up becoming a communication device; and speech developed out of language as what some evolutionary biologists call a derivative *exaptation*. Such developments seem to be inevitable evolutionary results; but they are either by-products of some other development or an adaptation to circumstances.[58]

Exapted for communication, first in the form of speech and later of script, language enabled human beings to enhance the non-verbal capacity with which they were already endowed. On the other hand, speech came to be *exapted* for modelling and to function, therefore, as a *secondary modelling system*. Beyond increasing the capacity for communication, speech also increases the capacity for innovation and for the 'play of musement'. The plurality of languages and 'linguistic creativity' (Chomsky) testify to the capacity of language, understood as a primary modelling device, for producing numerous possible worlds.

Modelling Device and Iconicity: Mind as a Sign System

Sebeok believes that language as a modelling device relates *iconically* (by resemblance) to the universe it models. This statement connects him directly with Peirce and Jakobson, both of whom stressed the importance of iconic signs. An equally important connection can be made with Ludwig Wittgenstein's *Tractatus*, particularly with the notion of 'picturing'.

The iconic relation can also be further explained and analysed through the distinction, discussed above in relation to biology, between *analogy* and *homology* analysed by Rossi-Landi.[59]

Analogy is direct similarity among isolated and static objects. It unites that which is divided, or at least that which is not united in a necessary way. Homology reveals how what seems – or usually seems – to be divided is in reality genetically united. It recognises the original unity. In other words, it restores the genetic two to one: the real and original process consists in dividing one into two. This distinction between analogy and homology is congenial to the general orientation of Sebeok's own research, given its association with biology. His method is homological.

This approach to the relation between language and world also has implications for the theory of knowledge, for the study of cognitive processes and psychology, which Sebeok directly addresses in terms of psycholinguistics and psychosemiotics. Relating semiotics to neuro-biology, he describes the mind as a sign system or model representing what is commonly called the surrounding world (*Umwelt*). This model is an icon, a kind of diagram where the most pertinent relations are of a spatial and temporal order. These relations are not fixed once and for all but can be fixed, modified and fixed again in correspondence (a resemblance relation) with the *Innenwelt* (inner world) of the human organism. On the basis of this model, comparable to a diagram or a map, the human mind may orientate itself by shifting from one node to another in the sign network, choosing each time the interpretive path considered most suitable.[60]

Semiotics of Life and Globalisation

Sebeok's semiotics of life may also provide an adequate understanding and comprehensive interpretation of 'globalisation'. Social production today is characterised by the automated industrial revolution, by global communication and by the global market. Not only does this imply expansion at a quantitative level, but also and above all a transformation at the level of quality, consisting in the fact that anything may be translated into commodities and that there is a continual production of new commodities. This has consequences for the very way people consider themselves. Generally, at the basis of all behaviour is the idea that what you produce is what you are, and that what you consume makes you the individual you want to be.

In the present age, communication is no longer just an intermediate phase in the production cycle (production, exchange, consumption), but has become the constitutive modality of production and consumption processes themselves. Not only is exchange itself communication, but also production and consumption must now be considered as communication. In fact, production is based on communicative processes whose most obvious expression is automation, robotisation, computerisation and online information. Thanks to the development of the means of communication, it is possible today for production to use (low cost) labour from the other side of the world. Consumption is ever more the consumption of commodities of a particular kind (that is to say, consumption of *messages* – telephone, television and telematic consumption, access to data bases, to educational and training processes and so on). Consequently, the whole production cycle has become communication and, consequently, this particular phase in social production may be characterised as the 'communication-production' phase.

Communication-production communicates the world as it is today. It is *global* communication, not only in the sense that it has expanded over the whole planet, but also that it now constitutes the world as it is. It may be better to say: it is communication *of* this world. Communication and reality, communication and being coincide. Thus, realistic politics (but if it is not realistic, it is not politics) is politics appropriate to global communication, to the modality of being of communication-production.

One of the risks involved in global communication-production is the risk of destroying communication itself. Here, the concept of destruction is not merely referred to that relatively simple or banal phenomenon commonly identified in literature and in filmic discourse as 'incommunicability' (a subjective-individualistic malady caused by the transition in communication to its current phase of development and inseparable from production). On the contrary, when we speak of the risk of the end of communication, we are referring to nothing less than the possibility of the *end of life* on the planet Earth. In this context of discourse, communication is obviously not understood in the reductive terms described above but rather is equated to life itself.

According to this broad interpretation, communication and life coincide, as Sebeok's semiotics in particular has made clear. From this point of view, the end of communication would in fact involve the end of life. And, indeed, production in today's society, unlike all other preceding phases in social development, is endowed with an enormous potential for *destruction*.[61]

For an adequate understanding of communication in its current historico-social specification as a worldwide phenomenon, as well as in its relationship with life over the whole planet (and remembering, therefore, that life and communication coincide), semiotics must adopt Sebeok's planetary perspective in both a spatial and temporal sense. Such an orientation will necessitate a more distanced approach to contemporary life that does not remain imprisoned within the confines of contemporary life itself.

With the spread of 'bio-power' (Michel Foucault)[62] and the policing of the body within the production process, world communication goes hand in hand with the spread of the concept of the individual as a separate and self-sufficient entity. The body has been understood and experienced as an isolated biological entity, as belonging to the individual, to the individual's sphere of belonging. Such an attitude has effaced cultural practices and has led to the marginalisation of world views based on the body as a site of social practice. The body as former 'social site' has been relegated to archaeology studied by folklore analysts, preserved in ethnological museums and in the histories of national literature.

The body was once perceived very differently in the popular culture of Europe, as Mikhail Bakhtin[63] reminds us in his study on medieval and Renaissance 'groteque realism'. Corporeal life was not conceived apart from the rest of terrestial existence but as a whole with other animal and plant life forms. Signs of the *grotesque* body (of which only very weak traces have survived in the present day) include ritual and carnival masks employed in popular festivities. 'Grotesque realism' in medieval popular culture (which pre-exists the various forms of bodily individualism connected with the rise of the mercantile bourgeoisie) presents the body as something not confined to itself but in symbiosis with other bodies, in relations of transformation and renewal that exceed the limits of individual life. In complete contrast, world communication today does not bring human bodies into the kinds of transformative relations experienced in medieval culture. It does not weaken the individualistic, private and static conception of the body; instead, it reinforces it.

As Michel Foucault in particular has revealed (but also recalling Rossi-Landi's acute analyses already in the 1970s),[64] specialist divisions between the sciences are crucial to the ideologico-social necessities of the 'new canon of the individualised body' (Bakhtin).[65] This, in turn, functions to assist the controlled insertion of bodies into the reproduction cycle of today's global production system.

This production system now requires social labour that is 'social' only to the extent that it is the average or sum of the work of individuals kept separate from and reciprocally indifferent to each other.

These labourers are not united by a common end, unless it be that of reproducing the production cycle, an end external to them and with respect to which they are alienated.

A global and detotalising approach to semiotics demands a high degree of awareness and acceptance of the other, a readiness to listen to others in their otherness, a capacity for opening to the other, to be measured not only in quantitative terms but also qualitatively. All semiotic interpretations by the student of signs, especially at a metasemiotic level, cannot take place without a relationship of dialogue with the other. 'Dialogism' is, in fact, a fundamental condition for a global approach to semiotics where to be orientated *globally* means actually to privilege the opening towards the particular and the local, rather than the tendency to totalise and enclose. This is an approach to the universe of signs that privileges the movement towards detotalisation rather than totalisation.

A Dialogic Approach to European and American Semiotics

In *Semiotics in the United States*, Sebeok analyses American semiotics at three different levels, at once closely interrelated and yet easily identifiable.

At the *first* level, he conducts a systematic and historical survey of the various theoretical trends, perspectives, problems, fields, specialisations and institutions that characterise American semiotics. Regarding the historical development, Sebeok assumes the difficult task of reconstructing the origins of American semiotics, identifying research couched in discourse that was not yet connoted as 'semiotics' at the time and that, in certain cases, is still today considered only marginally associated with semiotics or even completely distant from it.

The *second* level is theoretical and critical. Sebeok takes a stand with respect to given problems in semiotics: these include problems of a general order concerning, for instance, the delimitation of the field of semiotics or the construction of a general sign model. They also include problems of a more specific order concerning the various sectors and subsectors of the science, or 'doctrine of signs'. The impression that Sebeok gives is that the level of problems sets the agenda for the whole volume: it provides a fuller, more comprehensive picture of the first level and avoids limiting the volume to pure historical descriptivism.

The *third* level is connected with the second in the sense that, while developing and illustrating his theoretical views, Sebeok colours them with personal overtones and most often with amusing biographical anecdotes. Sebeok usually figures as one of the characters populating the stories, episodes and enterprises forming his narration. This is largely due to his surprising and perhaps unprecedented involvement in the organisation and promotion of the semiotic science at a world level – a cause to which he has been committed since the gradual emergence of semiotics as a discipline in its own right. Sebeok has been in direct contact with many of the authors mentioned in his volume and has many 'memories' of personal experiences with them. Consequently, these memories have found their way into his description of the problems and orientations characterising the semiotic globe.

With reference to these three shaping factors, another book by Sebeok similar to *Semiotics in the United States* is *The Sign & Its Masters*. Here too the historical, theoretico-critical and anecdoctal threads of Sebeok's discourse converge and interweave even more than in his other books, though the autobiographical aspect is never lacking in any one of them. *Semiotics in the United States* may also be related to *I Think I Am a Verb* in which autobiographical motivations are not lacking in the choice of topics, authors and personalities cited – including the eighteenth President of the United States of America, Ulysses S. Grant, whose words inspired the title of the volume.

What is immediately striking about Sebeok's work may be described as his 'dialogic' and 'polyphonic' approach (in the Bakhtinian sense of these words). Sebeok promotes dialogue among signs, among the different orders of signs, among different interpretive practices, domains and fields, as well as among the 'masters' of signs, including those who had never previously been in direct contact with each other or who did not even suspect they were dealing with signs (his so-called 'cryptosemioticians').

In line with his recognition of the importance of 'dialogism' for the development of thought, and even more broadly for the evolution of life generally, of which human thought processes are a part, Peirce too (forced into isolation after being excluded from academic life) had had occasion to write (in a letter to Victoria Lady Welby, 2 December 1904, and very much in accord with her own views) that:

[A]*fter all, philosophy can only be passed from mouth to mouth, where there is opportunity to object and cross-question.*[66]

As testified by his long teaching career and constant commitment to promoting the 'community of inquirers', for Sebeok the continuity of dialogic exchange is nothing less than of vital importance. Indeed, as Iris Smith states in her introduction to Sebeok's book of 1991, *American Signatures: Semiotic Inquiry and Method*, his own peculiar way of living his condition as an intellectual testifies to the fact that individual reflection must be measured against the reflections of others.

Semiosis Beyond Gaia?

The semiotic field extends over all terrestrial biological systems, from the sphere of molecular mechanisms at the lower limit, to a hypothetical entity at the upper limit christened 'Gaia', the Greek for 'Mother Earth' – a term introduced by scientists towards the end of the 1970s to designate the whole terrestrial ecosystem that englobes the interactive activity of different forms of life on Earth.[67]

As Sebeok says, alluding to the fantastic worlds of *Gulliver's Travels*, semiosis spreads over the *Lilliputian world* of molecular genetics and virology to *Gulliver's man-size world*, and finally to the *world of Brobdingnag*, Gaia, our gigantic bio-geo-chemical ecosystem.

And beyond? Can we assert that semiosis extends beyond Gaia? A 'beyond' understood in terms of space, but also of time? Is semiosis possible beyond Gaia, outside it, and beyond this gigantic organism's life span? Sebeok ponders this question too.[68]

With his research Sebeok takes stock of the impressive general progress and expansion of the field of semiotics during the past twenty to thirty years or so. Starting from a reductive definition of semiotics as the study of the exchange of any kind of message and related sign systems (which we have seen he criticises), he theorises semiotics as the 'play of musement' mediating between reality and illusion.

The 'play of musement' activating Sebeok's research is so free from prejudice that on examining the correspondence between life and semiosis, he even goes so far as to risk the hypothesis that the end of life does not necessarily imply the end of semiosis. With some probability, sign processes building limitless interpretants may continue in machines independently of humans. This Orwellian conclusion (formulated by Sebeok in his important and oft-cited text 'Semiosis and Semiotics: What Lies in their Future?') plays on the hypothesis of the machine as the sole place to remain for the workings of the 'life of signs'. In whatever way we wish to play on the words 'life' and 'signs', this conclusion proposes a sort of dystopia or negative utopia. From one viewpoint, the human can potentially become a form of non-life; as such, it would exclude the presence of signs according to Sebeok's formula: life = sign activity.

À propos the relation between life, semiosis and semiotics, and as a conclusion to this essay dedicated to Sebeok as a *Festschrift* on his eightieth birthday, we wish that the semiosis of Sebeok the man and the semiotics of Sebeok the semiotician may yet live a long and signifying life!

Notes

1. Thomas A. Sebeok was born in Budapest on 9 November 1920. He migrated to the United States of America in 1937 and became a citizen in 1944. He has been a faculty member of Indiana University since 1944 and is general editor of the journal of the International Association for Semiotic Studies, *Semiotica*, founded in Paris in 1969. Sebeok is among the figures who have most contributed to the institutionalisation of semiotics internationally, and to its configuration as 'global semiotics'.

2. See *The Sign & Its Masters*, Austin: University of Texas Press, 1979; second edition with a new Foreword by the author and Preface by J. Deely, Lanham: University Press of America, 1989.

3. For one of his most recent statements on this aspect, see Sebeok, 'Global Semiotics', plenary lecture delivered on 18 June 1994 as Honorary President of the Fifth Congress of the International Association for Semiotic Studies, held at the University of California, Berkeley. Now in Sebeok, *Global Semiotics*, Bloomington: Indiana University Press, 2001.

4. Sebeok, 'Semiotic and Communication: A Dialogue with Thomas A. Sebeok' (1987), in J. Y. Switzer et al. (eds), *The Southern Communication Journal 55*, 1990, p. 391.

5. Sebeok, *American Signatures: Semiotic Inquiry and Method*, intro. and ed. by I. Smith, Norman: University of Oklahoma Press, 1991, p. 124.

6. Sebeok, 'Global Semiotics', op. cit.

7. See Vladimir I. Vernadsky, *Biosfera*, Leningrad: Knizhnaia, 1926.

8. To gain a sense of the enterprise of global semiotics, see *Semiotik/Semiotics: A Handbook on the Sign-Theoretic Foundations of Nature and Culture*, ed. by R. Posner, K. Robering, T. A. Sebeok (three volumes), Berlin and New York: Walter de Gruyter, 1997, 1998; vol. 3 is forthcoming.

9. See Sebeok, 'Global Semiotics', op. cit. Interestingly enough, in 'Global Semiotics', a paper published almost 20 years after his book of 1976, *Contributions to the Doctrine of Signs* (Lisse: Peter de Ridder Press, 1976; second edition, Lanham: University Press of America, 1985), he no longer considers the debate on whether semiotics is a 'science', a 'theory' or a 'doctrine' of much consequence.

10. Sebeok, *Contributions to the Doctrine of Signs*, ibid., p. x.

11. Sebeok, *The Sign & Its Masters*, op. cit. See the programmatic chapters: 1, 'Semiosis in Nature and Culture', pp. 3–26; and 4, 'Ecumenicalism in Semiotics', pp. 61–84.

12. Sebeok, *The Play of Musement*, Bloomington:

Indiana University Press, 1981.

13. Umberto Eco, *Trattato di semiotica generale*, Milano: Bompiani, 1975; English translation, *A Theory of Semiotics*, Bloomington: Indiana University Press, 1976.

14. Sebeok, *I Think I Am a Verb: More Contributions to the Doctrine of Signs*, New York: Plenum Press, 1986; Sebeok, *Essays in Zoosemiotics*, ed. by M. Danesi, Toronto: University of Toronto Press, 1990; Sebeok, *A Sign is Just a Sign*, Bloomington: Indiana University Press, 1991; Sebeok, *American Signatures: Semiotic Inquiry and Method*, op. cit.; Sebeok, *Semiotics in the United States*, Bloomington and Indianapolis: Indiana University Press, 1991; Sebeok, *Signs: An Introduction to Semiotics*, Toronto: Toronto University Press, 1994; Sebeok, *Come comunicano gli animali che non parlano*, ed. by S. Petrilli, Bari: Edizioni dal Sud, 1998; Sebeok, *Global Semiotics*, op. cit.

15. Sebeok, *Perspectives in Zoosemiotics*, The Hague: Mouton, 1972; Sebeok (ed.), *Animal Communication: Techniques of Study and Results of Research*, Bloomington: Indiana University Press, 1968; Sebeok (ed.), *Sight, Sound, and Sense*, Bloomington and London: Indiana University Press, 1978; Sebeok (ed.), *How Animals Communicate*, Bloomington: Indiana University Press, 1979.

16. Charles K. Ogden and Ivor A. Richards, *The Meaning of Meaning: A Study of the Influence of Language upon Thought and of the Science of Symbolism*, London: Kegan Paul, 1923; new edition with introduction by U. Eco, New York: Harcourt Brace Janovich, 1989.

17. See Chapter 5, 'Vital Signs', in Sebeok, *I Think I Am a Verb*, op. cit.; see also the parts dedicated to these figures in Sebeok, *The Sign & Its Masters*, op. cit., and throughout Sebeok, *Semiotics in the United States*, op. cit.

18. Charles S. Peirce, *Collected Papers*, 8 vols., Cambridge (Mass.): The Belknap Press of Harvard University Press, 1931–58, vol. 5, § 119.

19. See Hans Blumenberg, *Die Lesbarkeit der Welt*, Frankfurt: Suhrkamp Verlag, 1981, p. 17.

20. Sebeok, 'Global Semiotics', op. cit.

21. See Sebeok (ed. with D. Jean Umiker-Sebeok), *Aboriginal Sign Languages of the Americas and Australia*, 2 vols., New York: Plenum Publishing Corporation, 1978.

22. See '"Talking" with Animals: Zoosemiotics Explained', in Sebeok, *The Play of Musement*, op. cit.

23. See Sebeok, *Signs*, op. cit., pp. 17–93.

24. See Sebeok's 'Preface' to *Contributions to the Doctrine of Signs*, op. cit.

25. Originally written on invitation from Norma Tasca,

representing the Associacao Portuguesa de Semiotica, for the Portuguese journal *Culture e Arte* 52, 1989; now available in *A Sign is Just a Sign*, op. cit., pp. 97–99.

26. Ibid., p. 97.

27. See *Semiotik/Semiotics*, vol. 1, Chapter III, *A Handbook on the Sign – Theoretic Foundations of Nature and Culture*, vol. 1, ed. by Roland Posner, Klaus Robering, Thomas A. Sebeok, Berlin and New York: Walter de Gruyter, 1997.

28. Sebeok's distinction between *language* and *speech* corresponds, if roughly, to the distinction between *Kognition* and *Sprache* drawn by Horst M. Muller in his book of 1987, *Evolution, Kognition und Sprache*, Berlin: Paul Parey. See ibid., p. 443.

29. Ibid., pp. 436–37.

30. In Sebeok, *The Sign & Its Masters*, op. cit.

31. See Mikhail M. Bakhtin (1965), *Rabelais and his World*, Cambridge: Massachusetts Institute of Technology, 1968.

32. See Sebeok's interesting considerations in 'The Semiotic Self', in Sebeok, *The Sign & Its Masters*, op. cit., Appendix I.

33. See Eco, *Trattato di semiotica generale*, op. cit.

34. See 'From Peirce (via Morris and Jakobson) to Sebeok: Interview with Thomas A. Sebeok', in Sebeok, *American Signatures: Semiotic Inquiry and Method*, op. cit., pp. 95–105.

35. Eco, *Trattato di semiotica generale*, op. cit.

36. Vailati's collected writings are now available in a work in three volumes: Vailati, *Scritti*, ed. by M. Quaranta, Sala Bolognese: Arnaldo Forni Editore, 1987.

37. See A. Ponzio, *Rossi-Landi e la filosofia del linguaggio*, Bari: Adriatica, 1988.

38. Sebeok, 'Peirce in Italia', *Alfabeta* 35, 28 April 1981.

39. See Sebeok, 'Looking in the Destination for What Should Have Been Sought in the Source', in Sebeok, *The Sign & Its Masters*, op. cit., pp. 85–106.

40. See 'Averse Stance', in Sebeok, *I Think I Am a Verb*, op. cit., pp. 145–48.

41. See Sebeok, 'Can Animals lie?', in Sebeok, *I Think I Am a Verb*, op. cit., pp. 126–30.

42. Examined under certain aspects by Morris, for example, in 'Mysticism and Its Language', 1957, a rather unusual paper for those who identify his work with his books of 1938 and 1946. *Foundations of the Theory of Signs* and *Signs, Language and Behavior*, now in Morris, *Writings on the General Theory of Signs*, T. A. Sebeok (ed.), The Hague and Paris: Mouton, 1971.

43. See Sebeok (with J. Umiker-Sebeok), '"You Know My Method." – A Juxtaposition of Charles S. Peirce and Sherlock Holmes', in Sebeok, *The Play of Musement*, op. cit.

44. Sebeok, 'Vital Signs', in Sebeok, *I Think I Am a Verb*, op. cit, pp. 77–78.

45. In Sebeok, *The Sign & Its Masters*, op. cit., p. 175.

46. S. Freud, *The Interpretation of Dreams* (1899), in *The Standard Edition of the Complete Psychological Works of Sigmund Freud*, vol. 4, London: Hogarth, 1953–74.

47. On these aspects, see in particular Sebeok's *The Play of Musement*, op. cit.

48. See Sebeok, *A Sign is Just a Sign*, op. cit., pp. 49–58, 68–82, and Sebeok, *Signs*, op. cit., pp. 117–27.

49. See Sebeok and Marcel Danesi, *The Forms of Meanings: Modeling Systems Theory and Semiotic Analysis*, Berlin and New York: Mouton de Gruyer, 2000, pp. 1–43.

50. Ibid., p. 5.

51. Ibid., pp. 44–45.

52. Ibid., pp. 82–85.

53. Ibid., pp. 120–29.

54. An exception is offered by a book by Giorgio Fano (1885–1963) entitled, *Origini e natura del linguaggio*, Torino: Einaudi, 1972; English translation and introduction by S. Petrilli, *Origins and Nature of Language*, Bloomington: Indiana University Press, 1992. See also the essays in Paul Cobley (ed.), *The*

Routledge Companion to Semiotics and Linguistics, London: Routledge, 2001.

55. 'From Peirce (via Morris and Jakobson) to Sebeok: Interview with Thomas A. Sebeok', in Sebeok, *American Signatures: Semiotic Inquiry and Method*, op. cit.

56. 'Communication, Language, and Speech: Evolutionary Considerations', included in Sebeok's book of 1986, *I Think I Am a Verb*, op. cit., pp. 10–16.

57. See Sebeok, 'Language as a Primary Modeling System', in Sebeok, *Signs*, op. cit., p. 125.

58. See S.J. Gould and E.S. Vrba, 'Exaptation: A Missing Term in the Science of Form', *Paleobiology* 8, 1982, pp. 4–15 and S.J. Gould and R. Lewontin, 'The Spandrels of San Marco', *Proceedings of the Royal Society* B, 205, 1979, pp. 581–98.

59. See Ferruccio Rossi-Landi, *Linguistics and Economics*, The Hague: Mouton, 1977, p. 75. See also Rossi-Landi, *Language as Work and Trade* (1968), English translation by M. Adams et al., South Hadley (Mass.): Bergin and Garvey, 1983.

60. See Sebeok, 'Naming in Animals with Reference to Playing: A Hypothesis', in Sebeok, *I Think I Am a Verb*, op. cit., Chapter 7.

61. See Augusto Ponzio and Susan Petrilli, *Il sentire nella comunicazione globale*, Rome: Meltemi, 2000.

62. See Luther H. Martin, Huch Gutman, Patrik H.

Hutton (eds), *Technologies of the Self: Seminar with Michel Foucault*, Amherst: The University of Massachusetts Press, 1988.

63. See Mikhail Bakhtin, *Problems of Dostoevsky's Poetics* (1963), Manchester: Manchester University Press, 1984; and Bakhtin, *Rabelais and His World*, op. cit.

64. See Rossi-Landi, *Language as Work and Trade*, op. cit.; Rossi-Landi, *Linguistics and Economics*, The Hague: Mouton, 1977; and Rossi-Landi, *Between Signs and Non-Signs*, ed. by S. Petrilli, Amsterdam: John Benjamins, 1992.

65. Bakhtin (1965), *Rabelais and his World*, op cit.

66. In Charles S. Hardwick (ed. and Introduction ix–xxxiv), *Semiotic and Significs: The Correspondence Between Charles S. Peirce and Victoria Lady Welby*, Bloomington and London: Indiana University Press, 1977.

67. See James E. Lovelock, *Gaia: A New Look at Life on Earth*, Oxford: Oxford University Press, 1979, and Lynn Margulis, *The Symbiotic Planet: A New Look at Evolution*, London: Phoenix Books, 1999.

68. See 'Semiosis and Semiotics: What Lies in their Future?', op. cit.

Further Reading

Eugen Baer, 'Thomas A. Sebeok's Doctrine of Signs', in Martin Krampen et al. (eds), *Classics of Semiotics*, New York: Plenum Press, 1987, pp. 181–210.

Jeff Bernard, 'Thomas A. Sebeok und die Zeichen des Lebens', in Peter Weibel (ed.), *Jenseits von Kunst* (= *Passagen Kunst*), Wien: Passagen Verlag, 1997, pp. 739–40.

Paul Bouissac et al., *Iconicity: Essays on the Nature of Culture: Festschrift for Thomas A. Sebeok on his 65th Birthday*, 'Foreword' by Claude Lévi-Strauss, Tübingen: Stauffenburg Verlag, 1986.

Instituto de investigaciones humanísticas, Universidad Veracruzana (ed.), *Semiosis* 26–29, número dedicado a Thomas A. Sebeok por sus 70 años, 1992.

Marcel Danesi, *The Body in the Sign: Thomas A. Sebeok and Semiotics*, Toronto: Legas, 1998.

John Deely (ed.), *Thomas A. Sebeok: Bibliography: 1942–1995*, Bloomington, Indiana: Eurolingua, 1995.

John Deely, 'Thomas A. Sebeok', entry in Paul Bouissac (ed.), *Encyclopedia of Semiotics*, Oxford: Oxford University Press, 1998.

John Deely, *Basics of Semiotics*, Bloomington: Indiana University Press, 1999.

John Deely and Susan Petrilli, *Semiotics in the United States and Beyond: Problems, People, and Perspectives*, papers resulting from 6–10 July 1992 seminar occasioned by Sebeok's *Semiotics in the United States*, and held at the Centro di Semiotica e linguistica di Urbino (Italy), special issue, *Semiotica*, 97–3/4, 1993.

Susan Petrilli, 'Sebeok', in P. Cobley (ed.), *The Routledge Companion to Semiotics and Linguistics*, London: Routledge, 2001.

Iris Smith, 'Thomas A. Sebeok: "The Semiotic Self" in America', in *American Signatures: Semiotic Inquiry and Method*, Norman: University of Oklahoma Press, 1991, pp. 3–18.

'Symbolicity', papers from the International Semioticians' Conference in Honour of Thomas A. Sebeok's 70th Birthday, Budapest and Wien, 30 September to 4 October 1990. Sources in *Semiotics* 11, Lanham: University Press of America, 1993.

Eero Tarasti (ed.), *Commentationes in Honorem Thomas A. Sebeok Octogenarii A.D. MM Editae*, Imatra, Finland: International Semiotics Institute, 2000. Among the 13 chapters of this book, see especially Eero Tarasti's editorial preface, Marcel Danesi's 'The Biosemiotic Paradigm of Thomas A. Sebeok' and Solomon Marcus's 'The Sebeok Factor: The

Right Man at the Right Place at the Right Moment'.
Norma Tasca (ed.), *Ensaios em homenagem a/Essays in honor of Thomas A. Sebeok*, *Cruzeiro Semiotico*, *22/25*, 1995.

Acknowledgements

We would like to thank Paul Cobley for his editorial advice on this book.